NEXTGEN LEARNERS PRESENTS

Think Like a Brain!™ Collection

Hippocampus Hal & The Memory Maze

How Your Brain Makes and Stores Memories

BOOK 7

Think Like a Brain!™
Collection Overview

Think Like a Brain!™ is a wildly engaging, silly-smart children's book collection that invites kids to explore how their brains work from the inside out. Through playful characters, imaginative settings, and rhythm-rich storytelling, this collection teaches neuroscience, social-emotional learning (SEL), and self-understanding in ways children absorb without even realizing they are learning.

Instead of telling children how to behave or what to feel, Think Like a Brain!™ shows them why their brains respond the way they do. Each story transforms complex brain science into friendly characters and relatable moments, helping kids feel capable, curious, and empowered rather than confused or ashamed.

Designed for children first and embraced by parents and educators as preventative education, this collection supports:

- Emotional regulation and self-awareness
- Executive functioning skills
- Neurodivergent understanding and inclusion
- Brain-body connection and growth
- Compassion for self and others

With humor, heart, and factual accuracy at its core, Think Like a Brain!™ helps children build the skills many adults wish they had learned sooner, one playful page at a time.

One of the collection's most powerful features is how the characters become real-life reminders for children long after the story ends. Once kids learn who the characters are and what jobs they do, they can begin to recognize those same processes in themselves. A big feeling might become, "It looks like Amy is trying to tell you something," or a moment of pause might sound like, "What does your Fred think?" By naming experiences through familiar characters, children learn to notice what is happening inside their brains, reflect on it, and move forward more quickly with understanding rather than frustration.

Series 1:
The Brainy Bunch (Brain Anatomy)

Series Overview: Meet the brain's core parts as a friendly team. Kids learn what each part does and how they work together to help thinking, feeling, moving, and learning.

1. **The Brainy Bunch**
   Focus: Brain overview and teamwork
2. **Brainstem Benny & The Automatic Adventure**
   Focus: Breathing, heartbeat, survival basics
3. **Bella Cerebellum's Balance Ballet**
   Focus: Balance, coordination, practice
4. **Olly Occipital & The Lookout Tower**
   Focus: Vision, visual processing
5. **Tammy Temporal & The Soundy Sound Storm**
   Focus: Hearing, language, sound processing
6. **Perry the Parietal & The Body Map Bash**
   Focus: Touch, space, body awareness
7. **Hippocampus Hal & The Memory Maze**
   Focus: Memory, learning, recall
8. **Amy the Amygdala & The Overactive Alarm**
   Focus: Emotional alarm system
9. **Luna Lefty & Roger Righty's Mixed Up Masterpiece**
   Focus: Left/right hemispheres, teamwork
10. **Callie the Connector & The Brain Bridge**
    Focus: Corpus callosum, brain-to-brain communication

Series 2:
Cortex Castle (Frontal Lobe Executive Functions)

Series Overview: A lively castle of executive-function helpers. Kids practice planning, starting, stopping, focusing, flexible thinking, time skills, and self-talk, with Frontal Fred guiding thoughtful choices.

1. **Cortex Castle**
   Focus: Executive function overview
2. **Frontal Fred & The Decision Dilemma**
   Focus: Decision-making, prioritizing
3. **Impulsive Ian & The Button Box**
   Focus: Impulse control, pause
4. **Memory Molly & The Wandering Thoughts**
   Focus: Working memory, holding steps
5. **Focus Frannie & The Great Distraction Chase**
   Focus: Attention, refocusing
6. **Planella & The Checklist Parade**
   Focus: Planning, sequencing
7. **Flexi & The Changey-Changeroo Day**
   Focus: Cognitive flexibility
8. **Self-Talk Sally & The Thoughtful Echoes**
   Focus: Self-talk, emotional regulation
9. **Starter Stacy & Finisher Finn**
   Focus: Task initiation, task completion
10. M**ilo the Minute Keeper & Mae the Moment Seeker**
    Focus: Time awareness, pacing
11. **The Power of Pausing with Penny**
    Focus: Pause skills, regulation

Series 3:
Busy Brain Café (Neurodivergent Affirmations)

Series Overview: A welcoming café where different brain styles are celebrated. These stories affirm neurodivergent experiences while teaching self-understanding, communication, and compassion for others.

1. **Nathan Neuron & The Bouncy Brain**
   Focus: ADHD metaphor, movement, attention
2. **Sara Sprinkle the Sensa & The Super Sensitive Day**
   Focus: Sensory sensitivity, overload tools
3. **Overwhelmed Owen & The Senses Overload**
   Focus: Sensory overload, coping strategies
4. **Interoceptive Ivy & The Inside Signals**
   Focus: Interoception, body signals
5. **Zane's Zigzag Thoughts & The Path to Peace**
   Focus: Anxiety, grounding
6. **The Busy Brain Café**
   Focus: Neurodivergent inclusion, different processing styles
7. **The Listening Light**
   Focus: Nonverbal communication, understanding
8. **The Language of Love**
   Focus: Connection, communication, compassion
9. **Trying Theo & The How-To Hiccup**
   Focus: Persistence, frustration tolerance
10. **Drowsy Drew & The Nighttime Repair**
    Focus: Sleep, brain repair
11. **Burned-Out Bonnie & The Rebuild**
    Focus: Burnout, rest, recovery

Series 4:
Steady Systems Squad (Regulation & Growth)

Series Overview: A squad that explains how brains regulate, reset, rest, and grow stronger over time, including stress waves, sleep crews, sensory filtering, and rebuilding after hard days.

1. **The Steady Systems Squad**
   Focus: Regulation systems overview
2. **Timothy Thalamus & The Traffic Tower**
   Focus: Sensory filtering, attention routing
3. **Danny Dimmer & Wake-Up Wonda**
   Focus: Arousal and alertness
4. **The Start, Stop, and Switch Station**
   Focus: Initiation, inhibition, shifting
5. **The Night Shift & The Sleepy Crew**
   Focus: Sleep systems, restoration
6. **The Stress Wave & The Return to Calm**
   Focus: Stress response, recovery
7. **The Brain Growth Gathering**
   Focus: Brain growth and development
8. **Paul's Practice Path**
   Focus: Neuroplasticity
9. **Daisy's Daydream Network**
   Focus: Daydream network, imagination
10. **Remy's Rebuild Crew**
    Focus: Recovery and healing

NEXTGEN LEARNERS PRESENTS

Think Like a Brain!™ Collection

Hippocampus Hal & The Memory Maze

How Your Brain Makes and Stores Memories

BOOK 7

WRITTEN &
ILLUSTRATED BY:

MAISEL MCLAULA

NEXTGEN LEARNERS PRESS™
CINCINNATI, OHIO

This book is published by NextGen Learners Press, an imprint of NextGen Learners LLC.

Maisel McLaula is a pen name used by the author of this work.

First Edition: 2026
Printed in the United States of America
ISBN: 978-1-972860-06-9

This book is intended for educational and informational purposes only. It is not intended to replace professional medical, psychological, or therapeutic advice, diagnosis, or treatment.
Always seek the advice of a qualified professional with any questions regarding a medical or mental health condition.

Artwork and visual elements in this book were created using a combination of original artwork, with enhanced original direction, digital illustration tools, and AI-assisted creative technologies.

A Note for My Readers:

Hi friend,
Your brain is incredible. It thinks, feels, remembers, solves, dances, creates, and so much more!
This series is here to help you understand how your brain works, and to cheer you on as you grow your own thinking superpowers.

Every part of your brain has a job.
Sometimes they get mixed up. That's okay!

When you learn what each part does, you can work with your brain instead of feeling like it's working against you.
So read, giggle, learn, and maybe even draw your own Brainy Bunch when you're done.

I'm cheering you on with every rhyme.

-Maisel McLaula

THINK LIKE A BRAIN!™

BRAIN ANATOMY COLLECTION

EXPLORE THE CASTLE INSIDE YOUR HEAD! MEET EACH ROOM AND ITS BRILLIANT RESIDENT, FROM FRONTAL FRED TO BENNY THE BRAINSTEM. WHIMSY, WONDER, AND NEURONS AWAIT!

For Cooper & Logan,
Brothers with curious minds and remarkable memories. You gather learning with care, store it with patience, and recall it right when it's needed most. Some memories stay forever. You and your incredible family are one of mine.

And...

To the beautifully wired, the curiously wired, and the wonderfully neurodivergent minds, if you forget, you can still figure it out.

You are exactly what a great brain looks like.

In a tall, leaning library lined with old maps, memories, and signs,
Lived Hippocampus Hal in his memory mines.
He collected your moments and filed them with care,
But sometimes forgot which memories lived where.

He charted first birthdays and picnics with pie,
The name of your teacher, and color of the sky

But when asked for a fact or a face from last spring,
He would twirl in his maze and forget the whole thing.

His filing system used song, smell, and sight,
But sometimes his halls were too dim or too bright.

And when stress filled the castle or chaos arrived,
Hal's maze would get jumbled and thoughts barely survived.

He sighed and he grumbled, "I know it was here,
A math fact, a lunchbox, a dance from last year?"

Fred came to visit with charts in his hand,
He said, “Let’s get organized, and make memory bands.”

Fred gave him a toolkit with tabs, scents, and sound,
“Memories stick best when you anchor them down.

Repeat them aloud, draw a doodle or two,
And group things together to help them stick through."

Hal nodded and labeled with stickers and glue,
He made memory bins in orange, green, and blue.

He practiced retrieval like finding lost socks,
By tracing the path, unlocking each box.

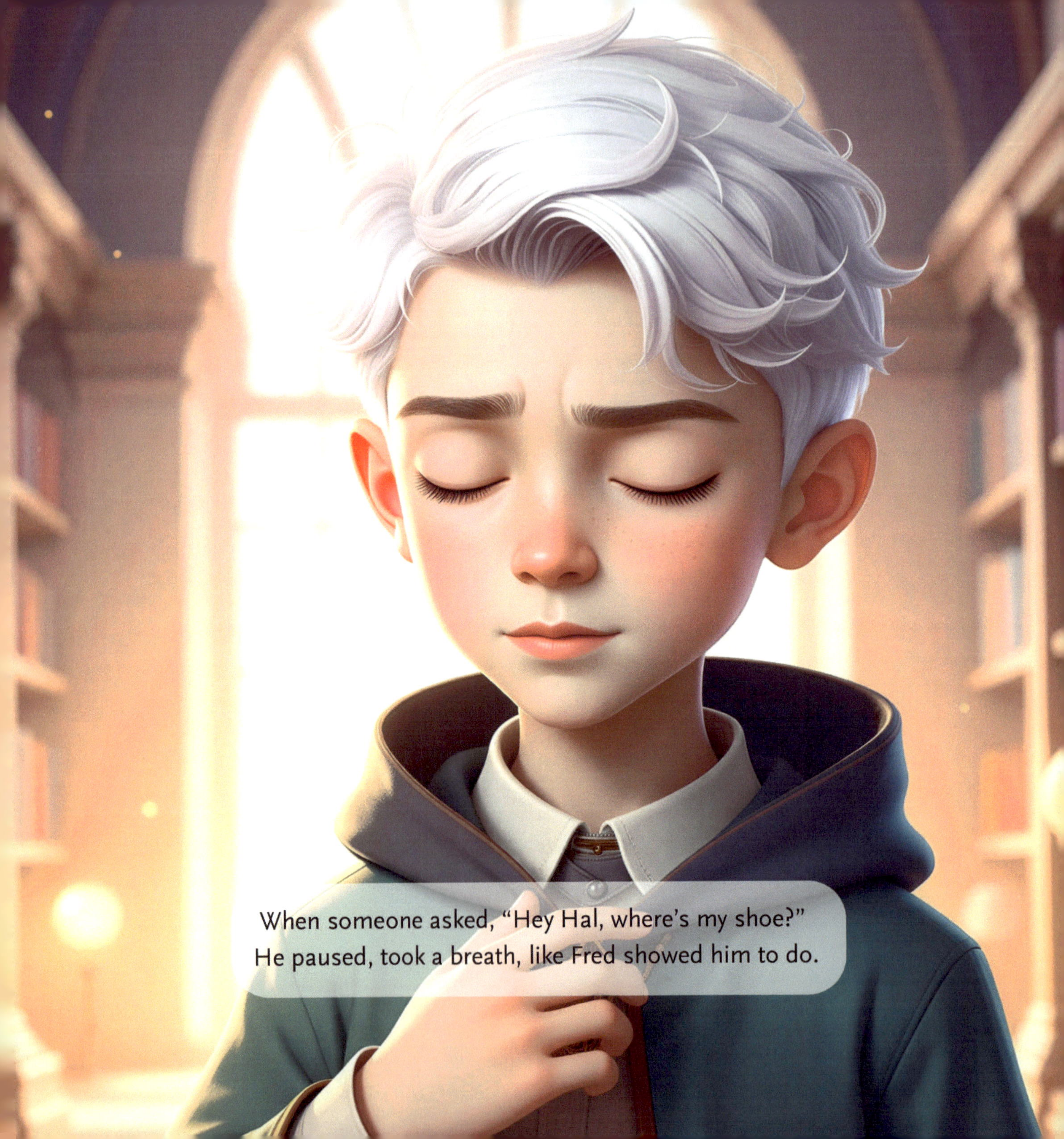

When someone asked, “Hey Hal, where’s my shoe?”
He paused, took a breath, like Fred showed him to do.

He pictured the morning, the busy bustle, and floor,
Then smiled, "It's waiting right by the front door!"

The Brainy Bunch clapped when Hal found old things,
Like birthdays and stories and favorite songs to sing.

His maze grew more friendly with glow paths and light,
And Hal felt much braver by day and by night.

He taught others tips in remembering things too,
Like drawing, repeating, or clapping a clue.

"You don't have to panic when memory's hazy,
You just need a system when details get crazy."

Now Hal is still Hal, he forgets now and then,
But he knows how to find those thoughts once again.

With anchors and maps and a journal or rhyme,
He retrieves and recalls most of the time.

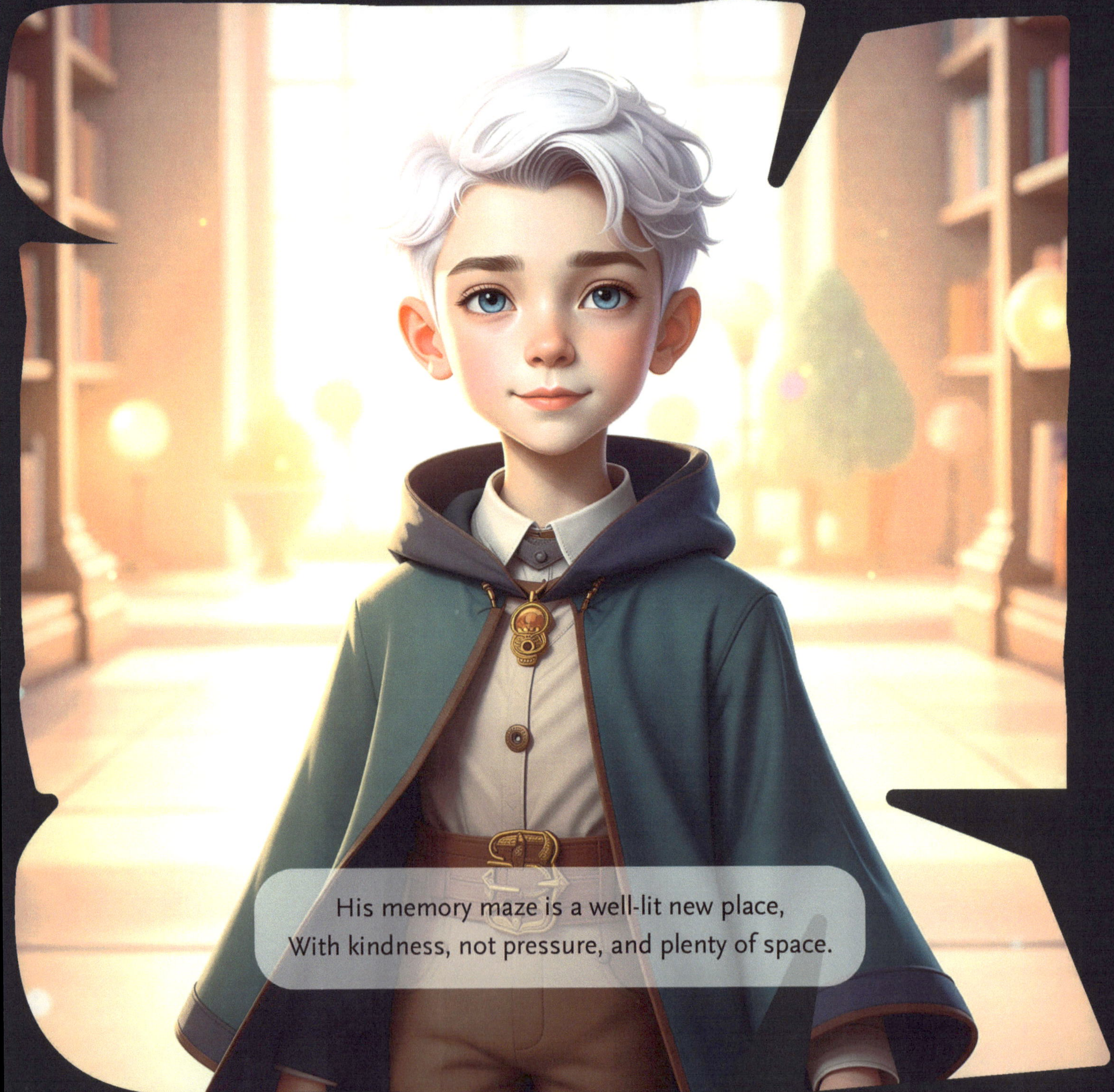
His memory maze is a well-lit new place,
With kindness, not pressure, and plenty of space.

He cheers when he finds what was hidden before,
And adds a new label to each memory door.

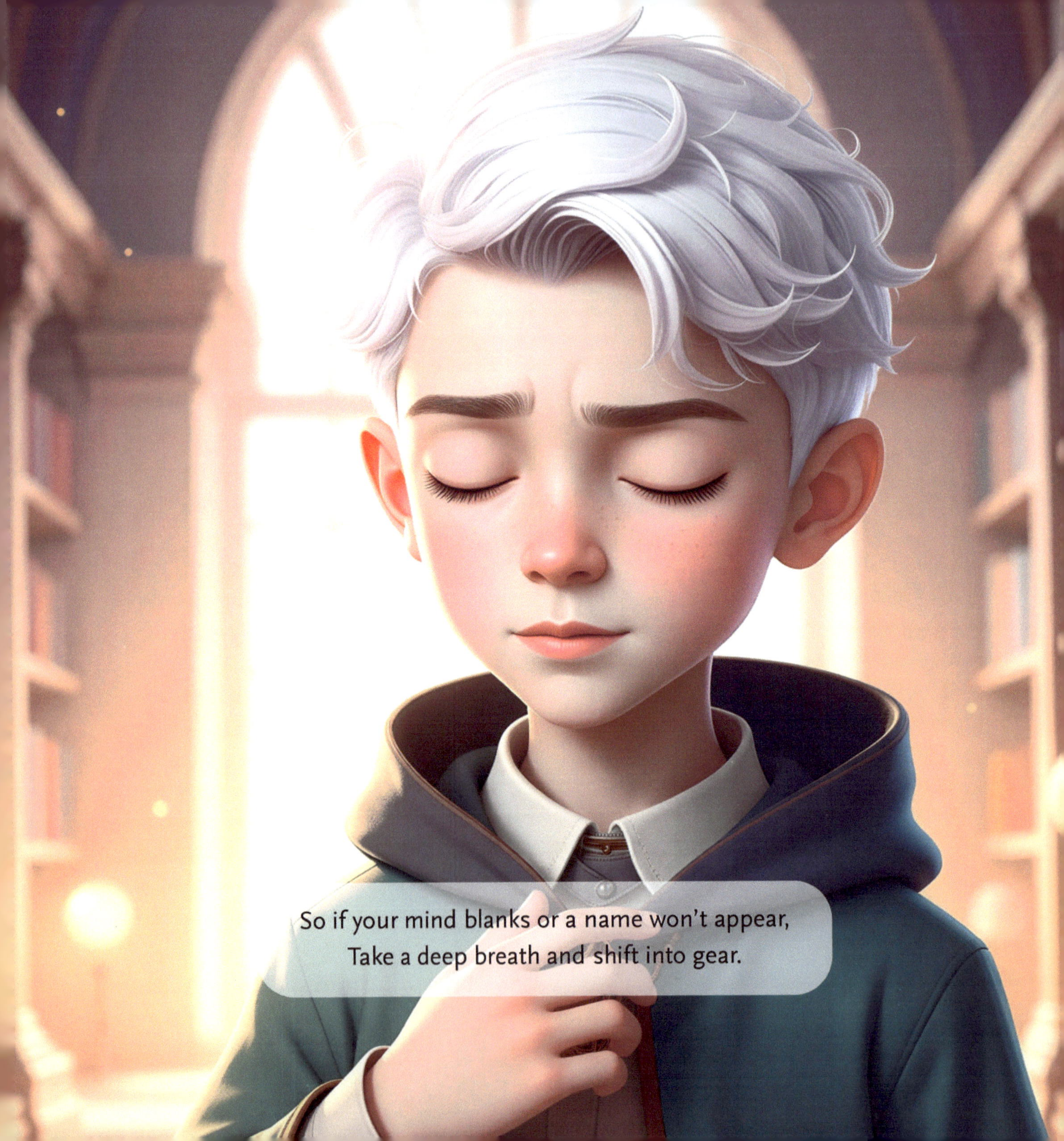

So if your mind blanks or a name won't appear,
Take a deep breath and shift into gear.

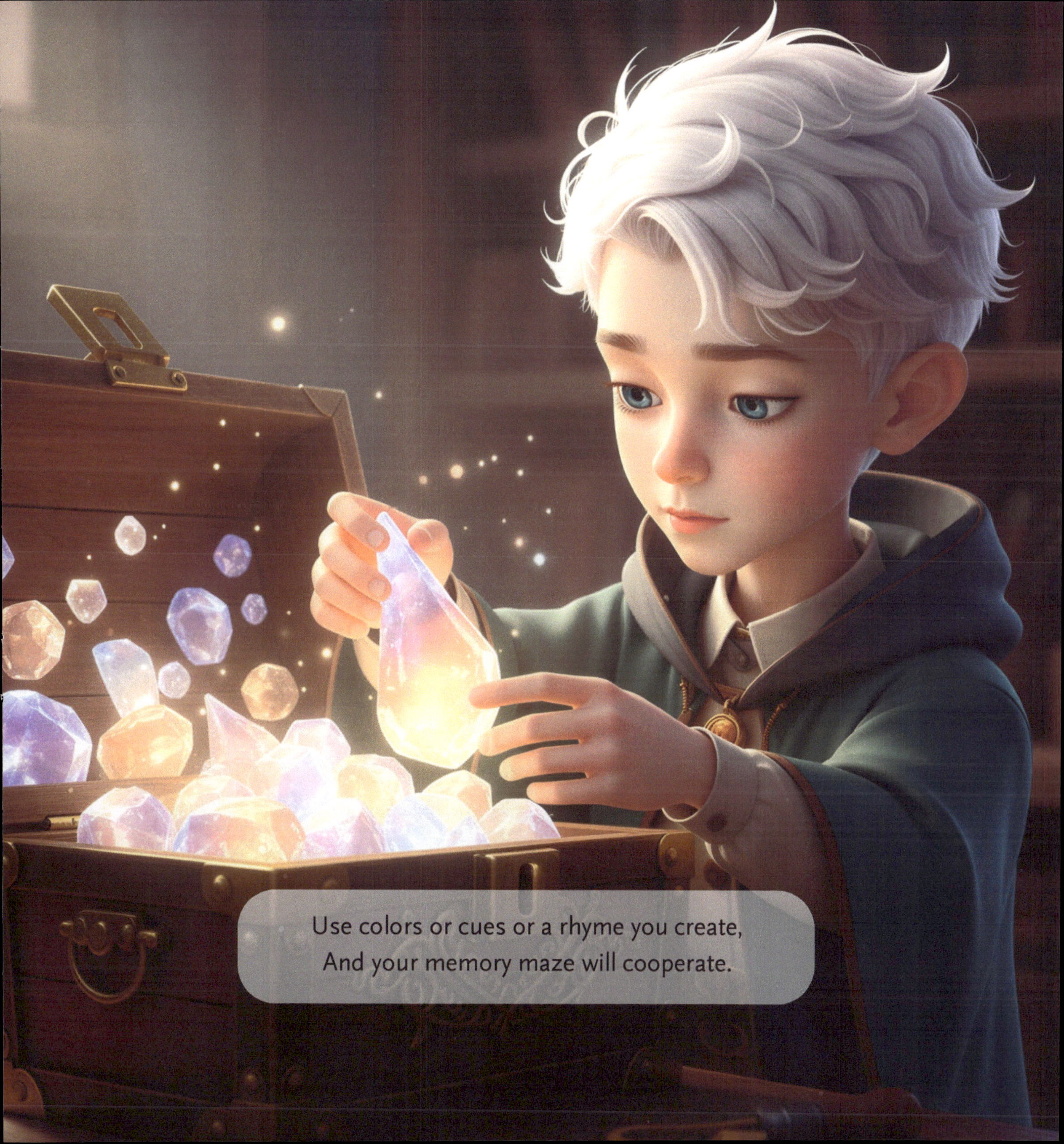

Use colors or cues or a rhyme you create,
And your memory maze will cooperate.

Hal’s Memory Tools

- Say it out loud and repeat with a beat
- Draw it, smell it, or clap out your feet
- Sort facts by theme, color, or zone
- Keep a journal or chart of your own

What Is the Hippocampus?

The hippocampus helps your brain make memories last.
It helps you remember things from the present and past.
When emotions are high or distractions are strong,
Your hippocampus might take a little bit long.

Try It Yourself!

- Draw your own memory maze with twists and a track
- Add recall signs so you can find your way back
- Practice remembering with games or a song
- Teach someone else so it lasts nice and long

Congratulations!

You've wandered the maze and found Hal inside,
With patience and tools, you learned how to guide.
Now memory feels like a path you can take,
To remember, recall, and avoid the mistake!

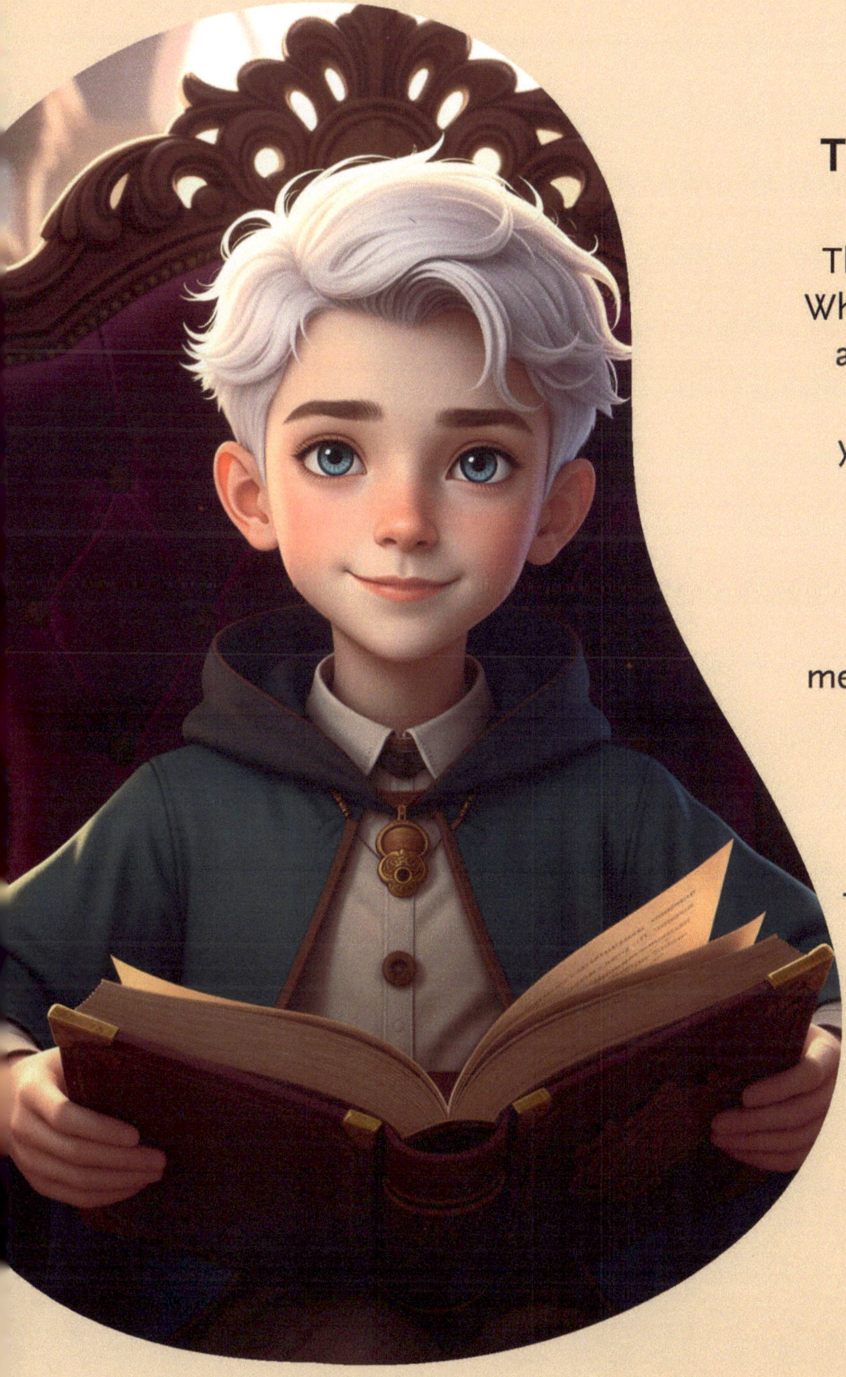

Thank You to My Readers

Thank you for reading this book.
Whether you're a parent, a teacher,
a therapist, or a brilliant young
brain explorer,
your interest in understanding
how the mind works
means the world.

Sharing these books
means spreading compassion and
curiosity.

You are the reason
these stories come alive.
Thank you for believing in the
power of playful learning.

Stay curious. Stay kind.
Stay brainy!

Hippocampus Hal & The Memory Maze
Teacher & Parent Guide

💡 What This Story Teaches:

• Brain part: Hippocampus (memory formation, learning, recall)

• Core idea: Memories stick best when they are organized, repeated, and connected to feelings or senses.

• Skills supported: working toward long-term memory, learning strategies, recall, sequencing

• Big message: Forgetting happens when stress is high. Calm and cues help memories return.

Hal shows children that memory is not magic. It is a process the brain practices.

🧠 Key Concepts for Grown-Ups

Hal represents the hippocampus, a curved structure deep inside the brain that helps turn experiences into memories. It works closely with emotions, senses, and attention.

When children feel rushed, anxious, overwhelmed, or distracted, the hippocampus has a harder time storing information.

This may look like:

- Forgetting instructions
- Trouble recalling learned material
- Difficulty following multi-step directions
- Losing track of belongings
- Remembering yesterday but not today

Important foundations:

- Memory improves when information is repeated and grouped.
- Sensory cues (visuals, sounds, movement) strengthen recall.
- Stress blocks memory pathways.
- Teaching others helps memories stick.

This story helps children understand that:

- Forgetting does not mean they are not smart.
- Memories return with patience and tools.
- Calm helps learning.

🗣️ Discussion Starters

- What made Hal's maze feel confusing?
- What tools helped Hal remember again?
- What helps you remember things?
- When does remembering feel hard?
- Can you think of something you practiced until it stuck?

Encourage children to share strategies,
not compare abilities.

👪 Try This Together:

1. Say It, Draw It

Say new information out loud, then draw a picture of it.

2. Memory Groups

Sort items into small groups (colors, types, categories).

3. Teach It Back

Let children explain what they learned to someone else.

4. Memory Walk

Walk while repeating facts or spelling words.

5. Recall Routines

Review learning at the same time each day.

🎯 Bonus Challenge

Create Hal's Memory Maze

Draw a simple maze and hide words, numbers, or pictures inside.

Ask:

- What clues help you find your way?
- What makes remembering easier?

Optional Labels:

- Recall Corner
- Practice Path
- Memory Gate

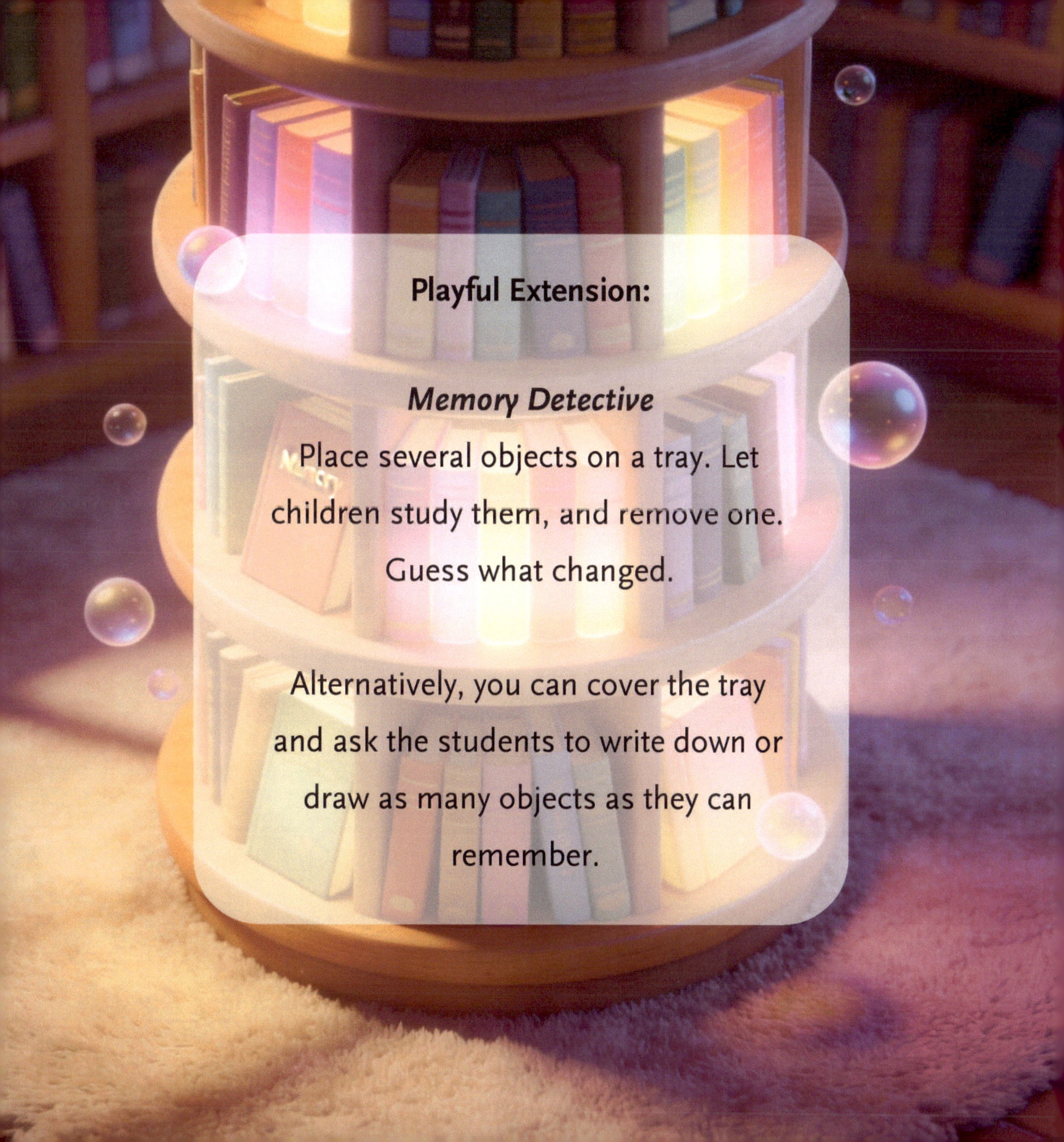

Playful Extension:

Memory Detective

Place several objects on a tray. Let children study them, and remove one. Guess what changed.

Alternatively, you can cover the tray and ask the students to write down or draw as many objects as they can remember.

🛠 Hal's Memory Toolbox:

Use these supports when children struggle with remembering:

- Break information into small chunks
- Use visuals and gestures
- Repeat instructions slowly
- Add movement to learning
- Reduce pressure during recall
- Build predictable routines
- Connect learning to emotion or story

Helpful phrases:

- "Your hippocampus is building paths."
- "Let's try it another way."
- "Calm helps memory."

Avoid framing forgetfulness as laziness. Memory grows with practice.

❤️ Final Note for Grown-Ups

Hal reminds us that memory is shaped by safety, repetition, and connection.

Children remember best when learning feels calm, meaningful, and playful.

Every review strengthens pathways.
Every cue lights the maze.

With patience and support, memories find their way home.

Resources & GROWTH

Resources:

(Hippocampus + memory formation and storage)

- Cleveland Clinic. Hippocampus: function/location (memory and learning overview).
- NCBI Bookshelf (StatPearls). Neuroanatomy, Hippocampus.
- NINDS (NIH). Brain Basics: Know Your Brain / Brain Basics resources.
- Cleveland Clinic. Brain: parts and function (overview pages).
- CASEL. CASEL's SEL Framework (self-awareness, self-management, social awareness, relationship skills, responsible decision-making).
- CDC. Child Development: Positive Parenting Tips and developmental guidance by age.
- MedlinePlus (U.S. National Library of Medicine) (memory and the brain)
- Encyclopedia Britannica

Growth:

(If you or someone you know wants to strengthen memory or learning strategies)

- Harvard Center on the Developing Child
    - Look for: learning and memory
- Understood.org
    - Search: memory strategies

Acknowledgments:

To my students: past, present, and future:
You taught me more than any textbook ever could. By watching how you explored, questioned, struggled, and discovered, I learned how minds truly grow. Every "aha" moment in these pages began with you.

To the parents, teachers, and guides walking beside children every day:
Thank you for listening, for advocating when things feel complicated, and for believing that understanding a child's inner world truly matters. Your patience and care shape futures in ways you may never fully see.

To my husband, my best friend, and my constant partner in learning:
Thank you for walking beside me, believing in this work, and holding space for creativity to exist. Sharing curiosity, growth, and creation with you has been one of the greatest honors of my life.

To the great thinkers who shaped how I see the world: including Dr. Seuss, Albert Einstein, Leonardo da Vinci, Stephen Hawking, Ludwig van Beethoven, just to name a few, and my English teacher, Mrs. Godfrey:
Your curiosity, courage, and imagination reminded me that learning is limitless, creativity is powerful, and questions matter. Mrs. Godfrey, your encouragement, belief in me, and love of words sparked my writing journey. If not for you, I may never have picked up a pen.

And to my publisher, NextGen Learners™:
It was by fate that we met, and by inspiration that we now move forward. Together, we build something meaningful. Together, we leave this legacy.

Lastly, to my own brain:
Complex, resilient, ever-adapting, and endlessly learning. Thank you for carrying stories, holding wonder, and reminding me that growth is always possible.

Maisel McLaula is the author of the Think Like a Brain!™ children's book collection, a story-first series that blends neuroscience, imagination, and emotional safety for young readers.

Drawing from decades of experience supporting children and families, her work centers on a simple but powerful belief: children deserve to understand and accept who they are, including how their brains work and why they make the choices they do.

Through playful characters and gentle metaphors, Maisel's stories help children discover the "why" behind their thoughts, feelings, and actions, turning confusion into clarity and self-judgment into self-compassion. By understanding their inner systems, children gain the power to respond with kindness toward themselves and to recognize those same patterns in others.

Maisel believes that when children feel safe inside their own minds, empathy grows naturally. The Think Like a Brain!™ collection offers a foundation for confidence, understanding, and care that grows with the child.

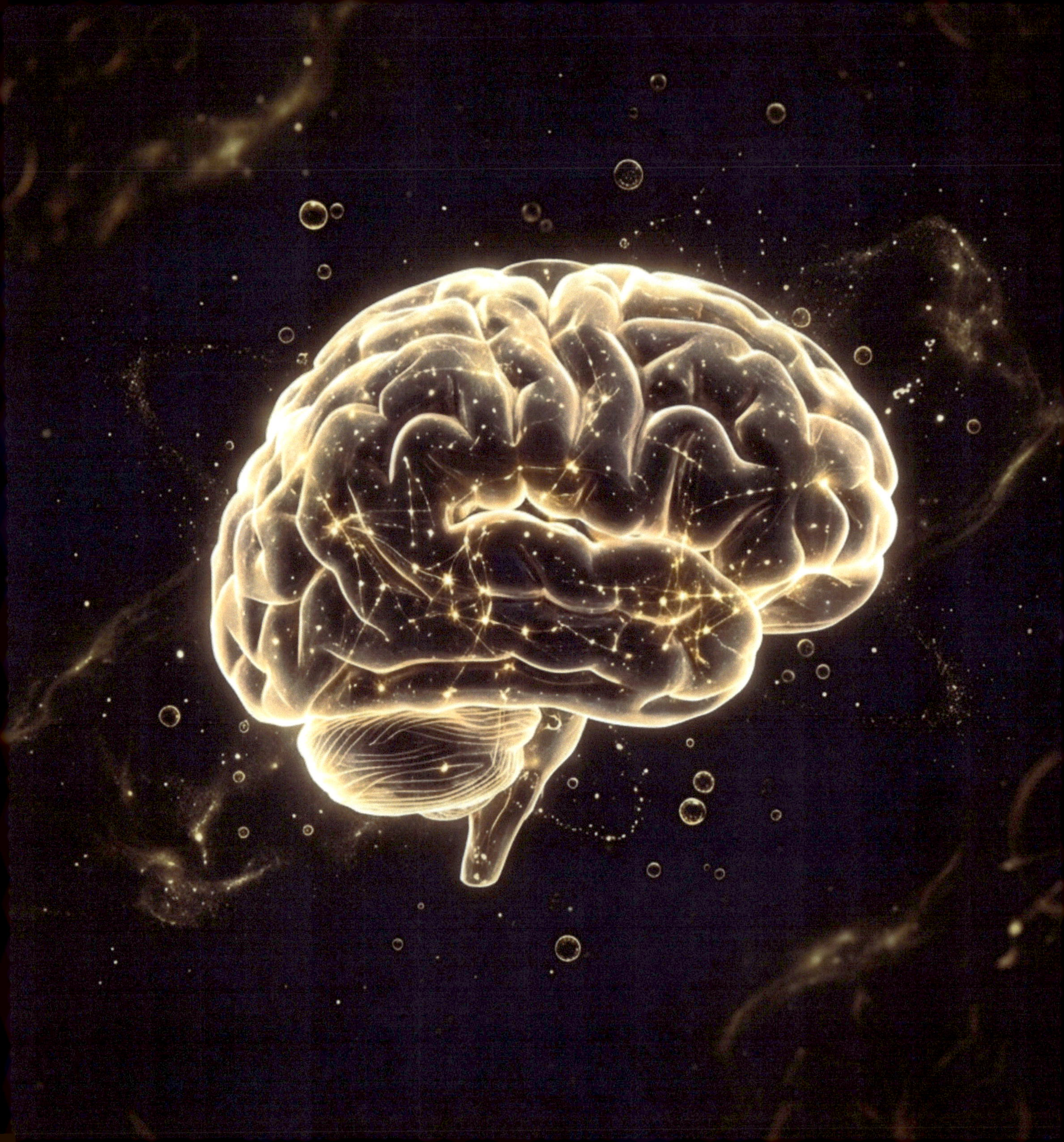

ABOUT THE PUBLISHER

NextGen Learners LLC is an educational publishing company dedicated to helping children understand how they think, feel, learn, and grow.

Through imaginative storytelling, neuroscience-informed concepts, and real-world application, NextGen Learners™ creates books and resources that support emotional development, critical thinking, self-awareness, and confidence in young minds.

Our work is designed not only for children, but for the parents, educators, caregivers, and professionals who walk beside them. Each book serves as a bridge, helping adults better understand the child in front of them, while helping children better understand themselves.

By creating shared language and meaningful connection points, our stories foster deeper relationships, reduce misunderstanding, and encourage acceptance, both within oneself and toward others.

We believe every child's mind is uniquely wired, and that understanding those differences is the foundation for compassion, growth, and lifelong learning.
With multiple collections designed to support cognitive, emotional, and developmental growth, NextGen Learners provides tools that children can carry with them, and tools that adults can use to guide them with clarity, patience, and purpose.

NG
NEXTGEN
LEARNERS
PRESS

MORE TO EXPLORE

The Brainy Bunch is just the beginning.
More stories, tools, and brain-friendly resources are on the way! Designed to help children, parents, and educators better understand how the brain works in everyday life.

FREE PRINTABLES

We've created printable pages to go along with this book, including:

-Reading Guides
-Brain helper visuals
-Character Cards

These are available for free as part of our growing resource library.

Our full website experience is built to include:

-Downloadable activity packs
-Parent & teacher guides
-Expanded learning tools
-Additional collections and topics

STAY CONNECTED

We love hearing how children connect with these stories!

If there is a topic, feeling, or challenge you would like to see explored in future books, we would love to hear from you!

Contact us: Inspire@NextGenLearners.com
NextGenLearners.com

Your ideas help shape future stories and support the growth of this collection.

This collection grows with you. New resources and stories are added regularly.

Maisel McLaula loves hearing from the tiny humans who read her books.

Did a story make you giggle, spark a big idea, or inspire you to draw your very own character? She would be delighted to see it! Send your letters and artwork to:

NextGen Learners
Attn: Maisel McLaula
8666 Beechmont Ave #1063
Cincinnati, OH 45255

Every letter is read with a smile, and your artwork just might make her whole day.

OPEN BRAIN

www.ingramcontent.com/pod-product-compliance
Lightning Source LLC
Chambersburg PA
CBHW041322210926
51847CB00002B/25
* 9 7 8 1 9 7 2 8 6 0 0 6 9 *